YOUR KNOWLEDGE HAS VALUE

- We will publish your bachelor's and master's thesis, essays and papers

- Your own eBook and book - sold worldwide in all relevant shops

- Earn money with each sale

Upload your text at www.GRIN.com and publish for free

Siyuan Chen

The theory and discovery of the Higgs boson

GRIN Verlag

Bibliografische Information der Deutschen Nationalbibliothek:

Die Deutsche Bibliothek verzeichnet diese Publikation in der Deutschen National-
bibliografie; detaillierte bibliografische Daten sind im Internet über http://dnb.d-
nb.de/ abrufbar.

Imprint:

Copyright © 2013 GRIN Verlag GmbH
Druck und Bindung: Books on Demand GmbH, Norderstedt Germany
ISBN: 978-3-656-74526-6

The theory and discovery of the Higgs boson

Siyuan Chen

12th December, 2013

King's College London
6CCP3132 Literature Review in Physics

Abstract

An overview of the steps that lead to the discovery of the Higgs boson is presented. Starting with the theoretical background framework, the Standard Model of particel physics, the Higgs field will be introduced as an addition. This extra field provides the mechanism for spontaneous symmetry breaking, that is needed to explain the existence of massive particles. An overview of the steps of the experimental search to the discovery of the Higgs boson is given in the second part of this article. Its mass has been measured to be $\sim 125.4 \pm 0.4(\text{stat}) \pm 0.5(\text{sys})$ GeV.

The Standard Model is briefly summarised. The Higgs mechanism is derived from an Abelian Model, applied to the gauge bosons of the electroweak model of Weinberg and Salam. A simple estimate of the Higgs mass is given by its derivation and the estimation of its self-coupling and vacuum expectation value.

Experimental results will be presented from the CMS and ATLAS detectors at the LHC, alongside with a description of the Large Hadron Collider at CERN and possible directions for future experiments beyond the Standard Model.

Contents

1 Standard Model of particle physics

1.1 History of the Standard Model

The Standard Model is a gauge quantum field theory and owes its modern form to many contributors in its history of development.

Quantum electrodynamics QED can be seen as the first Quantum field theory, it was popularised by Wolfgang Pauli in the 1940s. QED describes the electromagnetic field and explains its effect on charged quantum mechanical particles. In the light of QED Chen Ning Yang and Robert Mills proposed a non-Abelian gauge field theory in 1954 to describe the weak interaction[1].

In 1961 Sheldon Glashow published "Partial-symmetries of weak interactions" [2], in which he described a way to combine electromagnetic with weak interactions. Three years later François Englert and Robert Brout[3] and Peter Higgs[4] independently proposed a mechanism, that would allow symmetry breaking. It is commonly known as the Higgs mechanism today and was incorporated in 1967/1968 by Weinberg and Salam into Glashow's electroweak theory, forming the basis of the modern Standard Model[5][6]. The last major addition to the Standard Model was the inclusion of the strong interactions in the 1970s.

The Standard Model[7] consists of 16 elementary particles, which are divided into 3 groups: 6 Quarks, 6 Leptons and 4 Bosons/force carriers
The quark and leptons are also devided in 3 generations:
- first generation: Up quark, Down quark, electron, electron neutrino
- second generation: Charm quark, Strange quark, muon, muon neutrino
- third generation: Top quark, Bottom quark, tau, tau neutrino
The difference between corresponding particles in different generations is their mass energy. The higher the generation the heavier the particle is.

The quarks and leptons are the matter particles, that the universe is build of. They are usually also referred to as fermions. The bosons are the so-called force carriers, due to the fact that they mediate forces between fermions.

In addition to the 6 quarks, 6 leptons and 4 bosons, there exist their corresponding anti-matter particles. The antiparticle has the same mass as its matter counterpart, but opposite electromagnetic charge and other properties. In collisions of matter and antimatter, they will annihilate each other and create energy, in accordance to the energy-matter equivalence.

The Higgs boson is a unique particle of the Standard Model. Its importance is to give the other particles their mass.

1.2 Theoretical description

In the Standard Model[7] the elementary particles are described as quantised excitations of physical fields, it is thus a Quantum field theory. The particles can interact with each other via their fields and therefore form a dynamical

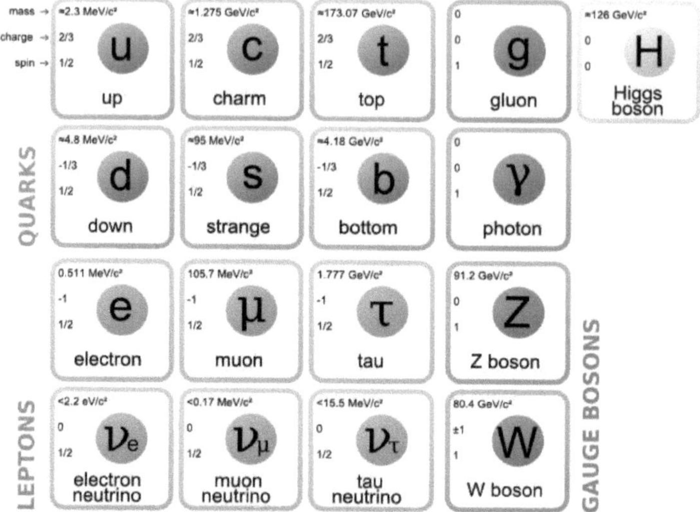

Figure 1: The 16 elementary particles and the Higgs boson[8]

system. A dynamical system is usually described by its kinetic energy $\mathfrak{T}$ and potential energy $\mathfrak{V}$, the quantity $\mathfrak{L} = \mathfrak{T} - \mathfrak{V}$ is called the Lagrangian. The Lagrangian of the Standard Model is[9]

$$\mathfrak{L} = -\frac{1}{4}F_{\mu v}F^{\mu v} + i\bar{\psi}\slashed{D}\psi + hc. + \psi_i y_{ij}\psi_j\phi + hc. + |D_\mu\phi|^2 - V(\phi) \qquad (1)$$

with

$$-\frac{1}{4}F_{\mu v}F^{\mu v} = -\frac{1}{2}tr(\mathbf{G}_{\mu v}\mathbf{G}^{\mu v}) - \frac{1}{8}tr(\mathbf{W}_{\mu v}\mathbf{W}^{\mu v}) - \frac{1}{4}B_{\mu v}B^{\mu v} \qquad (2)$$

The $hc.$ mean the corresponding Hermitian conjugates.

This is a very simplified version, with $\mathfrak{L}_f = i\bar{\psi}\slashed{D}\psi + hc.$ being the Lagrangian of the fermions, $\mathfrak{L}_Y = \psi_i y_{ij}\psi_j\phi + hc.$ being the Lagrangian of the Yukawa coupling and $-\frac{1}{4}F_{\mu v}F^{\mu v}$ being the gauge term of the bosons.

The bosons form a mathematical Lie-group[10] called SU(3)×SU(2)×U(1). A unitary group U(n) is a group of unitary matrices that have the size n×n. The group operation is the matrix multiplication. The inverse of a unitary matrix is its Hermitian conjugate.

The special unitary groups SU(n) are the subgroups of U(n), in which all element matrices have $detM = 1$.

The SU(3)×SU(2)×U(1) group has 8×3×1 generators respectively.

Mathematically, it is the direct product of three groups, which forms another group.

Physically, it can be interpreted as 8 gluons, 3 bosons (W^+, W^-, Z^0) and 1 photon γ.

The gauge bosons have properties that have to be conserved in interactions. A Quantum field theory, in which the Lagrangian of a system is invariant under certain symmetry gauge transformations, is called a Gauge theory.
Applying a gauge transformation on the Lagrangian and checking for its invariance will tell us if a physical quantity is conserved in the system. This is a result of Noether's theorems[11][12].

The gauge transforms that correlate to the conservation of electric charge can be written as

$$A_\mu(x) \to A_\mu(x) - \partial_\mu \eta(x) \tag{3}$$

and

$$\phi(x) \to e^{-ie\eta(x)}\phi(x) \tag{4}$$

One can see that a simple mass term $\frac{m^2}{2}A_\mu A^\mu$ for the Langrangian does not hold the gauge invariance and thus electric charge will not be conserved.

$$\frac{m^2}{2}A_\mu A^\mu \neq \frac{m^2}{2}(A_\mu A^\mu - A_\mu \partial^\mu \eta - A^\mu \partial_\mu \eta + \partial_\mu \eta \partial^\mu \eta) \tag{5}$$

Consequently, the theory dictates that the particles have to be massless. However, everyday experience tells us that there is mass. To solve this problem a mechanism has to be added into the Lagrangian, that will give mass to the particles. It is called the Higgs mechanism and will be introduced in section 2.

1.3 Physical interactions

We know 4 fundamental forces in nature: Gravity, Weak force, Strong force and the electromagnetic force
Excluding gravity, the Standard Model[7] describes the other 3 forces as force fields in space whose excitation would correspond to bosons. Those bosons can be seen as particles whch interacts between fermions.

The electromagnetic force is carried by the photon, the weak force is carried by the W and Z bosons and the stong force is carried by the gluons.

The fundamental interactions between bosons and fermions can be described by Feynman diagrams in figure 2:
- quark and anti-quark mediated by a gluon
- fermion and anti-fermion mediated by a Z-boson
- electron, muon, tau (electron generations e') and their corresponding anti-neutrino $\bar{\nu}'$ mediated by a W-boson
- Up, Charm, Top quark (Up quark generations u') and Down, Strange, Bottom quark (Down quark generations d') mediated by a W-boson
- charged fermions f' and their anti-particles $\bar{f}'$ mediated by a photon γ

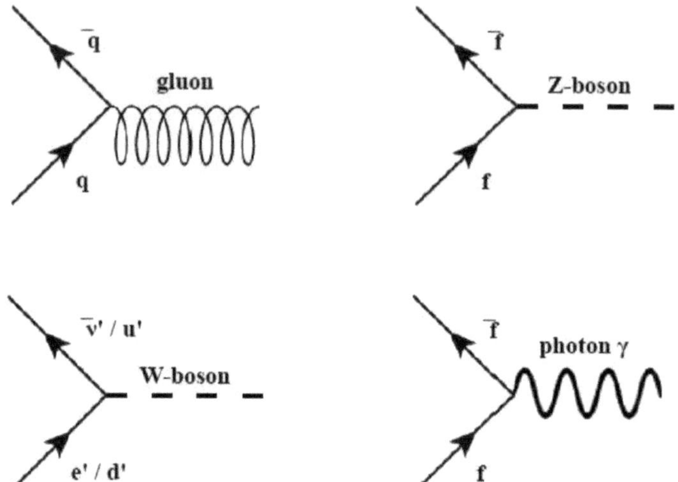

Figure 2: The possible interactions being bosons and fermions

2 Models of the Higgs mechanism

2.1 Abelian Higgs Model

In the Abelian Higgs model[13][14][15] the Lagrangian takes the form

$$\mathcal{L} = -\frac{1}{4}F_{\mu\nu}F^{\mu\nu} + |D_\mu\phi|^2 - V(\phi) \tag{6}$$

with

$$F_{\mu v} = \partial_v A_\mu - \partial_\mu A_v \tag{7}$$

$$D_\mu = \partial_\mu - ieA_\mu \tag{8}$$

$$V(\phi) = \mu^2 |\phi|^2 + \lambda(|\phi|^2)^2 \tag{9}$$

The first term describes the field strength of a simple gauge field of U(1), the second and third term add the kinetic and potential term of a complex scalar field.

D_μ is called the covariant derivative, μ and λ are constants and ϕ is the wave function associated to the particle.

In the U(1) symmetry the local gauge invariances that conserve electromagnetic charge are

$$A_\mu(x) \rightarrow A_\mu(x) - \partial_\mu\eta(x) \tag{10}$$

and

$$\phi(x) \to e^{-ie\eta(x)}\phi(x) \tag{11}$$

Inserting (10) into (8) and (11) into (9) will give

$$D'_\mu = \partial_\mu - ieA_\mu(x) - ie\partial_\mu\eta(x) \tag{12}$$

$$V(\phi') = \mu^2 \left| e^{-ie\eta(x)}\phi(x) \right|^2 + \lambda(\left| e^{-ie\eta(x)}\phi(x) \right|^2)^2 \tag{13}$$

Using $|\psi|^2 = \psi^\dagger\psi$, one can see that $V(\phi') = V(\phi)$ since

$$\left| e^{-ie\eta(x)}\phi(x) \right|^2 = e^{-ie\eta(x)\dagger}e^{-ie\eta(x)}|\phi(x)|^2 = |\phi(x)|^2 \tag{14}$$

Furthermore

$$D_\mu\phi' = \partial_\mu\phi' - ieA_\mu(x)\phi' - ie\partial_\mu(\eta(x))\phi' \tag{15}$$

$$= \partial_\mu(e^{ie\eta(x)}\phi(x)) - ieA_\mu(x)e^{ie\eta(x)}\phi - ie\partial_\mu(\eta(x))e^{ie\eta(x)}\phi \tag{16}$$

$$= \partial_\mu(\phi)e^{ie\eta(x)} + \phi e^{ie\eta(x)}ie\partial_\mu(\eta(x)) - ieA_\mu(x)e^{ie\eta(x)}\phi - ie\partial_\mu(\eta(x))e^{ie\eta(x)}\phi \tag{17}$$

The second term cancels with the fourth term, leaving

$$D_\mu\phi' = (\partial_\mu - ieA_\mu)e^{ie\eta(x)}\phi = (\partial_\mu - ieA_\mu)\phi' \tag{18}$$

Compared to

$$D_\mu\phi = (\partial_\mu - ieA_\mu)\phi \tag{19}$$

one can see the invariance of the covariant derivative under the two transformations $A_\mu \to A_\mu - \partial_\mu\eta(x)$ and $\phi \to e^{ie\eta(x)}\phi$.

Going back to (9), two cases can be distinguished: μ^2 is positive or negative. If it is positive, V will look like a parabola with a minimum at zero. The minimum is called the Vacuum expectation value(VEV) as it is the state of lowest energy for the system. A parabola-like shaped potential with the VEV being at zero does not break any symmetry.
If it is negative, V will look like what is called a mexican hat potential. The minimum of the potential can be calculated

$$\frac{dV}{d\phi} = 2\mu^2\phi + 4\lambda\phi^3 = 2\phi(\mu^2 + 2\lambda\phi^2) = 0 \tag{20}$$

As the potential shown below in figure 3, the $\phi = 0$ solution will lead to a local maximum. Thus leaving

$$-\mu^2 + 2\lambda\phi^2 = 0 \tag{21}$$

Solving for ϕ we derive at the VEV

$$<\phi> = \pm\sqrt{-\frac{\mu^2}{2\lambda}} \equiv \pm\frac{v}{\sqrt{2}} \tag{22}$$

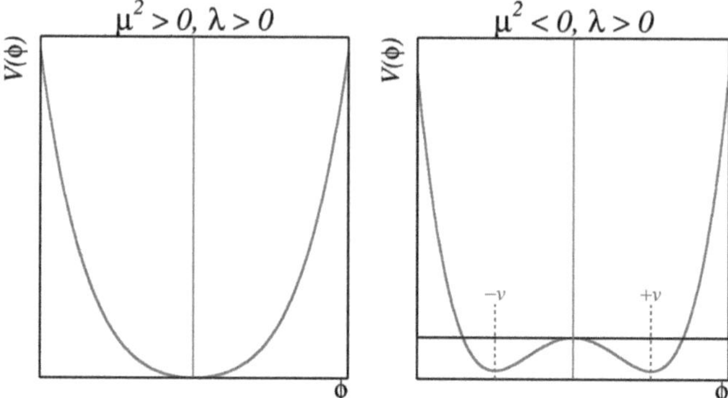

Figure 3: A 2-D comparison of V(ϕ) for $\mu^2 > 0$ and $\mu^2 < 0$[16]

Since there are two different ϕ with the lowest energy possible for the system, after some time has passed the system has to decide for one or the other, at this point the symmetry has been broken.
For convenience we will define ϕ as

$$\phi \equiv \frac{v+h}{\sqrt{2}} e^{i\frac{\chi}{v}} \tag{23}$$

with χ and h being real field with no VEV.
Substituing (23) back into the Lagrangian (6) we will get the following

$$\mathcal{L} = -\frac{1}{4}F_{\mu\nu}F^{\mu\nu} - evA_\mu\partial^\mu\chi + \frac{e^2v^2}{2}A_\mu A^\mu + \frac{1}{2}(\partial_\mu h\partial^\mu h - 2\mu^2 h^2) + \frac{1}{2}\partial_\mu\chi\partial^\mu\chi + (h, \chi \; interactions) \tag{24}$$

The Lagrangian (24) now describes a photon with mass $M_A = ev$, a scalar field h with mass $-\mu^2 > 0$, a massless field χ and a mixed term $(\chi - A_\mu)$-contribution, that can be removed via a gauge transformation

$$A_\mu \rightarrow A_\mu + \frac{1}{ev}\partial_\mu\chi \tag{25}$$

Then

$$\mathcal{L} = -\frac{1}{4}F_{\mu\nu}F^{\mu\nu} - evA'_\mu\partial^\mu\chi + \frac{e^2v^2}{2}A'_\mu A^{\mu\prime} + \frac{1}{2}(\partial_\mu h\partial^\mu h - 2\mu^2 h^2) + \frac{1}{2}\partial_\mu\chi\partial^\mu\chi \tag{26}$$

$$\mathcal{L} = -\frac{1}{4}F_{\mu\nu}F^{\mu\nu} - evA_\mu\partial^\mu\chi - \frac{1}{2}\partial_\mu\chi\partial^\mu\chi + \frac{e^2v^2}{2}A_\mu A^\mu + evA_\mu\partial^\mu\chi + \frac{1}{2}\partial_\mu\chi\partial^\mu\chi + \frac{1}{2}(\partial_\mu h\partial^\mu h - 2\mu^2 h^2) \tag{27}$$

$$\mathcal{L} = -\frac{1}{4}F_{\mu\nu}F^{\mu\nu} + \frac{e^2v^2}{2}A_\mu A^\mu + \frac{1}{2}(\partial_\mu h\partial^\mu h - 2\mu^2 h^2) \tag{28}$$

χ is usually called the Goldstone field and h the Higgs field. The disappearance of the Goldstone field from the Lagrangian is considered to give mass to the photon.[18] Finally, the Lagrangian (28) consists only of the gauge term, a photon field A_μ with mass term $\frac{ev^2}{2}$, and a scalar field h with mass term $-\mu^2 \equiv v^2\lambda$.

2.2 Weinberg-Salam Model

The Weinberg-Salam Model of electroweak interaction is described by a SU(2)$\times$U(1) gauge theory with three SU(2) bosons W_μ^i and one U(1) boson B_μ. With $i = 1, 2, 3$ being an index indicating the field.
The gauge terms of the Lagrangian[13][14][15] are

$$\mathfrak{L}_G = -\frac{1}{4}W_{\mu v}^i W^{\mu v i} - \frac{1}{4}B_{\mu v}B^{\mu v} \tag{29}$$

where

$$W_{\mu v}^i = \partial_v W_\mu^i - \partial_\mu W_v^i + g\epsilon^{ijk}W_\mu^j W_v^k \tag{30}$$

and

$$B_{\mu v} = \partial_v B_\mu - \partial_\mu B_v \tag{31}$$

The field is applied on a complex scalar SU(2)-doublet Φ

$$\Phi = \begin{pmatrix} \phi^+ \\ \phi^0 \end{pmatrix} \tag{32}$$

with a vacuum expectation value of

$$< \Phi >= \begin{pmatrix} 0 \\ \frac{v}{\sqrt{2}} \end{pmatrix} \tag{33}$$

The direction of the minimum is not determined in SU(2) space, therefore we can abitrarily chose the vacuum expectation value above as shown in figure 4
A speficic choice of the VEV, the weak hypercharge and the electromagnetic charge will allow the SU(2)$\times$U(1) to break into a U(1) space. Now it is possible to apply the mechanism described in section 2.1 to give masses to the gauge bosons.
The contribution of the scalar doublet to the Lagrangian is

$$\mathfrak{L}_H = (D^\mu\Phi)^\dagger(D_\mu\Phi) - V(\Phi) \tag{34}$$

with

$$D_\mu = \partial_\mu + i\frac{g}{2}\tau^i W_\mu^i + i\frac{g'}{2}B_\mu \tag{35}$$

and

$$V(\Phi) = \mu^2\left|\Phi^\dagger\Phi\right| + \lambda\left|\Phi^\dagger\Phi\right|^2 \tag{36}$$

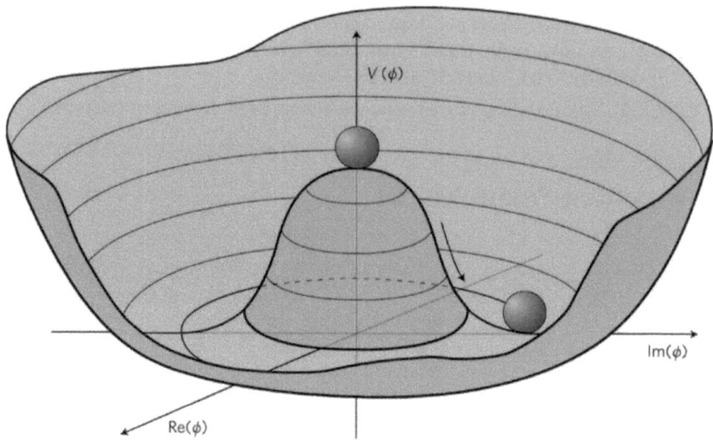

Figure 4: A 3-D plot of the potential V(Φ) for negative μ^2[19]

τ^i are the Pauli matrixes here. g and g' are the corresponding coupling constants to the SU(2) gauge field.

Just as before, we consider $\mu^2 < 0$, then the vacuum expectation value will not be at zero, but at $\begin{pmatrix} 0 \\ \frac{v}{\sqrt{2}} \end{pmatrix}$ (33). In the unitary gauge the Goldstone bosons are "eaten" by the other field, so that after symmetry breaking Φ can be written as

$$\Phi = \begin{pmatrix} 0 \\ \frac{v+h}{\sqrt{2}} \end{pmatrix} \tag{37}$$

Inserting this Φ into the combined Lagrangian of (29) and (34) gives

$$\mathcal{L} = -\frac{1}{4}W^i_{\mu v}W^{\mu v i} - \frac{1}{4}B_{\mu v}B^{\mu v} + (D^\mu\Phi)^\dagger(D_\mu\Phi) - V(\Phi) \tag{38}$$

$$\mathcal{L} = -\frac{1}{4}W^i_{\mu v}W^{\mu v i} - \frac{1}{4}B_{\mu v}B^{\mu v} + \left|(\partial_\mu + i\frac{g}{2}\tau^i W^i_\mu + i\frac{g'}{2}B_\mu)\begin{pmatrix} o \\ v \end{pmatrix}\right|^2 - V(\Phi) \tag{39}$$

The contribution to the masses of the gauge bosons is

$$\frac{1}{2}\left|(\frac{g}{2}\tau^i W^i_\mu + \frac{g'}{2}B_\mu)\begin{pmatrix} 0 \\ v \end{pmatrix}\right|^2 \tag{40}$$

$$= \frac{v^2}{8}\left|(g\tau^i W^i_\mu + g'B_\mu)\begin{pmatrix} 0 \\ 1 \end{pmatrix}\right|^2 \tag{41}$$

$$= \frac{v^2}{8}\left|\begin{pmatrix} gW^1_\mu - igW^2_\mu \\ -gW^3_\mu + g'B_\mu \end{pmatrix}\right|^2 \tag{42}$$

$$= \frac{v^2}{8}[g^2((W^1_\mu)^2 + (W^2_\mu)^2) + (gW^3_\mu - g'B_\mu)^2] \tag{43}$$

The physical fields $W^\pm$, Z^0 and A are defined as

$$W^\pm_\mu = \frac{1}{\sqrt{2}}(W^1_\mu \mp iW^2_\mu) \tag{44}$$

$$Z_\mu = \frac{-g'B_\mu + gW^3_\mu}{\sqrt{g^2 + g'^2}} \tag{45}$$

$$A_\mu = \frac{gB_\mu + g'W^3_\mu}{\sqrt{g^2 + g'^2}} \tag{46}$$

These definitions can be rewritten as

$$W^+_\mu W^{\mu-} = \frac{1}{2}((W^1_\mu)^2 + (W^2_\mu)^2) \tag{47}$$

$$Z^2_\mu(g^2 + g'^2) = (-g'B_\mu + gW^3_\mu)^2 \tag{48}$$

Inserting the definitions (46) (47) and (48) into the Lagrangian(43) gives

$$= \frac{v^2 g^2}{4}W^+_\mu W^{\mu-} + \frac{v^2(g^2 + g'^2)}{4}Z^2_\mu + 0 \cdot A^2_\mu \tag{49}$$

The masses of the bosons are obtained as

$$(M_{W^\pm})^2 = \frac{1}{4}g^2 v^2 \tag{50}$$

$$M^2_Z = \frac{1}{4}(g^2 + g'^2)v^2 \tag{51}$$

$$M_A = 0 \tag{52}$$

Furthermore the potential $V(\Phi)$ yields the mass of the Higgs field h

$$V(\Phi) = \mu^2(\frac{h}{\sqrt{2}})^2 + \lambda(\frac{h}{\sqrt{2}})^4 = -v^2\lambda(\frac{h}{\sqrt{2}})^2 + \lambda(\frac{h}{\sqrt{2}})^4 \tag{53}$$

$$= -\frac{v^2\lambda}{2}h^2 + \frac{\lambda}{4}h^4 = \frac{\mu^2}{2}h^2 + \frac{\lambda}{4}h^4 \tag{54}$$

From this, one can see the mass term $m^2_H = v^2\lambda = -\mu^2$, the self-coupling constant $\lambda/2$ and the VEV v of the Higgs field.

3 The Higgs boson

3.1 Mass of the Higgs boson

Even though no prediction on the mass of the Higgs boson is made by the Standard Model, there are other theoretical considerations which could give at least some hints and limits for its mass.

Knowing that the Higgs boson mass is[21]

$$m_H = \sqrt{\frac{\lambda}{2}} v \tag{55}$$

estimations on v and λ can be used to give a range for the Higgs mass. From experimental data the VEV of the Higgs field can be derived as $v = (\sqrt{2}G_F)^{-0.5} \approx 246$ GeV, λ the self-coupling constant of the Higgs field is to be estimated. Assuming that it is between 0 and 1, gives a range of the Higgs mass from 0 to 173 GeV.

A more theoretical approach can be done by considering the integral of the beta function. It describes the dependence of a coupling constant g for a specific physical process in QFT at a given energy scale Λ, it is defined as

$$\beta(g) = \frac{\partial g}{\partial log(\mu)} \tag{56}$$

μ is a renormalisation factor and typically taken to be $\frac{\Lambda^2}{v^2}$.

The lower limit can be estimated by arguments of vacuum stability[24]. The upper limit can be discussed by reasons of triviality[25]. Both limits can be computed with the pertubation theory.

Figure 5 shows a graphical result of the computation.

At low Λ the Higgs mass can vary from close to zero to nearly infinity. This is due to the fact that at very low energy scale, the coupling constant can become infinte, which then leads to an infinite integral and a possible infinite range for the Higss mass. It is also commonly known as Landau pole.

At high Λ the range for a stable Higgs potential field becomes very narrow. The upper limit decreases like $(log\frac{\Lambda^2}{v^2})^{-1}$, whereas the lower limit only gradually increases with $log\frac{\Lambda^2}{v^2}$.

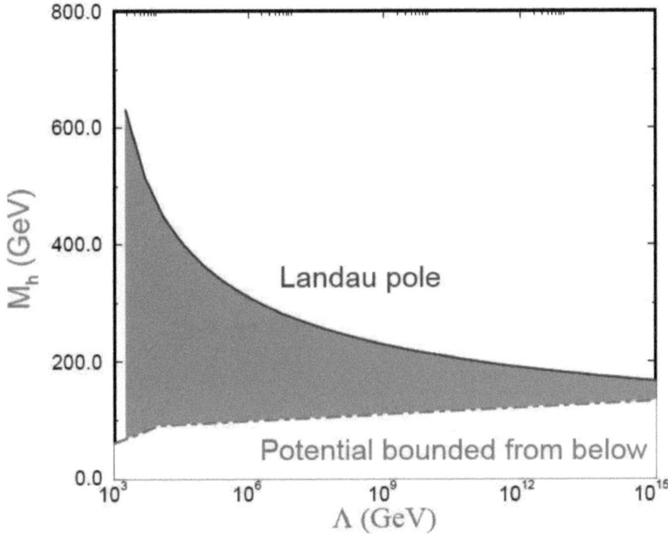

Figure 5: Limits for the Higgs mass computated from the beta function for the range 10^3 GeV to 10^{15} GeV[13]

3.2   Production and decay of the Higgs boson

The Higgs boson can be created in many different ways, 4 possible ways are:
- gluon-gluon fusion to a Higgs boson
- top quark and anti top quark fusion to a Higgs boson
- Higgs Strahlung: fermion and antifermion fuse to a virtual W/Z boson, which emits a Higgs boson
- fermion and antifermion interact through virtual W/Z bosons, which fuse to a Higgs boson
The Feynman diagrams are shown in figure 6.

The Higgs boson can decay in various ways, also called decay modes:
- H $\rightarrow$ ZZ* $\rightarrow$ 4 leptons
- H $\rightarrow$ $\gamma\gamma$
- H $\rightarrow$ WW* $\rightarrow$ electron + neutrino, muon + neutrino
- H $\rightarrow$ $\tau\tau$
- H $\rightarrow$ bb
The first two modes are pictured in figure 7.

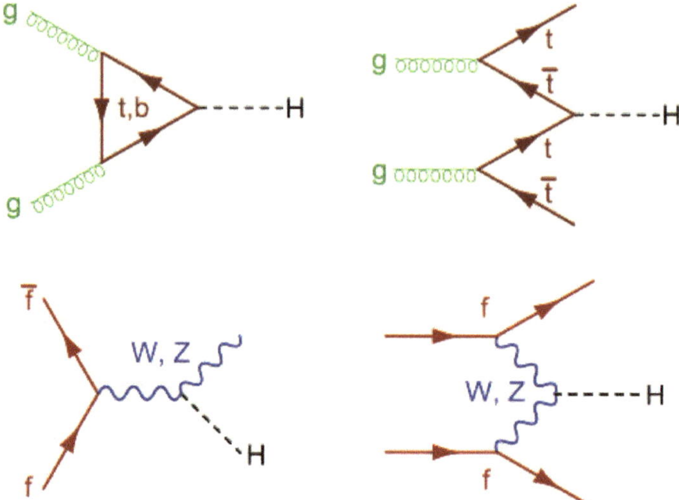

Figure 6: Feynman diagrams of four possible ways to produce a Higgs boson: left top gg-fusion, right top $t\bar{t}$-fusion, left bottom Higgs Strahlung, right bottom WZ-fusion

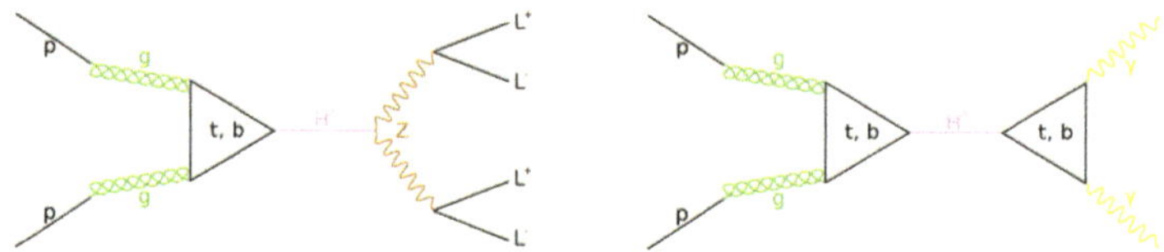

Figure 7: 2 Higgs decay modes: on the left $H \to ZZ \to 4l$, on the right $H \to \gamma\gamma$

The different probability for each mode as a function of the Higgs mass is called the Higgs branching ratio[23].
Depending on various factors the probability for a Higgs boson to decay into a specific mode is different. They are guided by the theoretical calculation on Feynman diagram of the interaction according to Fermi's Golden rule. The transition amplitude of a certain reaction can be calculated as a path integral over all possible ways for the reaction to happen from a given initial to a final state. The square of the transition amplitude is proportional to the probability of that reaction.

At lower mass the Higgs boson will primarily decay into a $b\bar{b}$ pair, at higher mass WW and ZZ decay will dominate. Thus, the branching ratio diagram can

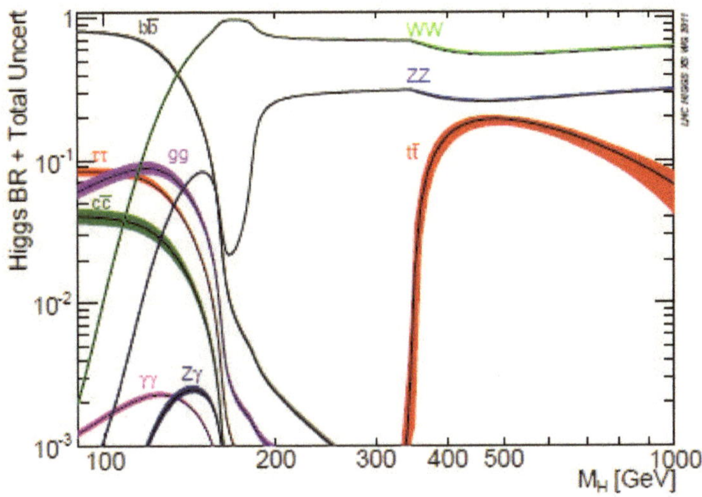

Figure 8: Higgs branching ratio in the range ~100 to 1000 GeV

be divided into two section: a higher mass and a lower mass section. The critical point is roughly at 160 GeV.

In the lower mass regime, bb decay will dominate, but many other decay processess are also possible, such as $\tau\tau$, cc, gg.

In the intermediate regime, between 100-160 GeV, the probability for lower mass decay processes falls sharply to approach zero. Two very minor decay processes emerge, that are $\gamma\gamma$ and $Z\gamma$. Those two decay processes are related to photons and even though the probability is only of the order 1 in 1000, photons can be detected quite well for the experimental search of the Higgs boson. Additionally, the WW and ZZ decay start to increase.

In the higher mass regime, the dominating decay processes are WW, ZZ and $t\bar{t}$. It should be noted that that the tt process is only relevant from 350 GeV on and reaches a maximum at roughly 500 GeV, before it starts to decrease in probability again. Since the WW and ZZ decay already have a significant probability between 100 and 200 GeV and will dominate for higher masses, they will be important decay processes to look at.

4 Experimental Search

4.1 History of events

In the 1990s the first large-scale experimental search for the existence of the Higgs boson was executed at the Large Electron-Positron Collider LEP at CERN.

The LEP focussed on the creation of a Higgs from the electron-positron interaction. The fact that no observations were made on the Z boson decaying or radiating a Higgs boson suggested that if a Higgs boson existed it had to be heavier than the Z boson and thus its mass energy equivalent must be higher than 91 GeV. With more collision data recorded and higher energies reached by accelerating the electron-positron pair, it was concluded that the mass of the Higgs boson has to be larger than 114.4 GeV.[26] In November 2000 the LEP was closed down to be replaced by the LHC.

The efforts of Tevatron at the Fermilab near Chicago, USA, in the time between the LEP and the LHC, should also be noted. The results from the proton-antiproton collider suggested a exclusion of 147-180 GeV for the Higgs mass as well as an excess of events between 115-135 GeV for futher research.[27]

At CERN the LEP was suceeded by the Large Hadron Collider LHC. In 2008/9 the LHC started its search for the Higgs boson. Operating at energies around 7-8 TeV, collision data has been collected and analysed. On the 4th of July 2012 a new unknown Higgs-like boson was announced to be found at 125-6 GeV with a significance of 5.0 σ. This announcement was based on the analysis of two independently working collaborations and detectors ATLAS[28] and CMS[29], both measuring an excess of events around 124 to 126 GeV. Further analysis of the properties of that new found boson were checked against the theoretical predictions of the Standard Model for the Higgs boson. It was confirmed by CERN in March 2013 that this boson is likely to have spin-0 and a positive parity, both requirements for the Higgs boson are met. Alongside with its interaction with other particels it is strongly suggested that it is indeed the Higgs boson.[32][33][34]

On the 8th October 2013 the Nobel Prize Committee of the Royal Swedish Academy of Sciences announced that P. Higgs and F. Englert will be awarded the Nobel Prize in Physics of 2013 "for the theoretical discovery of a mechanism that contributes to our understanding of the origin of mass of subatomic particles, and which recently was confirmed through the discovery of the predicted fundamental particle, by the ATLAS and CMS experiments at CERN's Large Hadron Collider".[35]

4.2 The Large Hadron Collider LHC

The complex system of accelerators and detectors at CERN[36] are used for many different experiments, which are aimed to help to better understand the physics of the early universe. Amongst those experiments, the focus of this article is mainly on those that are related to explore the Standard Model of

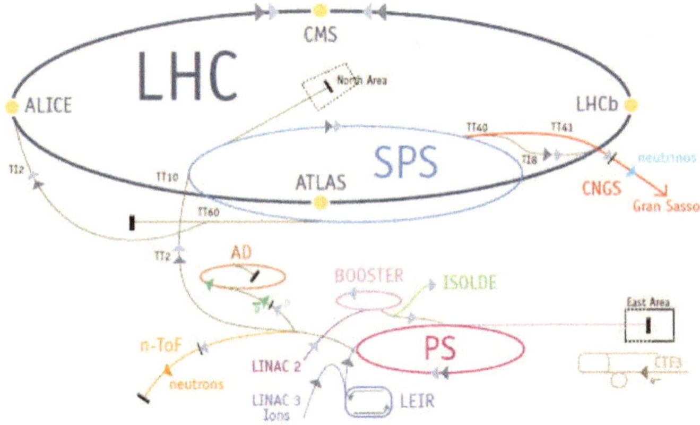

Figure 9: Scheme of the LHC with all accelerators and detectors

Particle Physics and especially ATLAS and CMS, which are dedicated to look not only for the Higgs boson, but also for physics beyond the Standard Model.

As the name Large Hadron Collider suggests, it is an accelerator and collider of hadrons. Hadrons are particles that are purely built of quarks, the proton or the neutron are examples of hadrons. In the LHC, protons are accelerated up to 99.9999991% of the speed of light and then collided head on head. The collison will produce many other particles that then will be detected and analysed.

The series consists of 5 successive accelerators: Linac 2, Proton Synchroton Booster PSB, Proton Syncrotron PS, Super Proton Synchrotron SPS, and finally the Large Hadron Collider LHC. The proton are injected at 50 MeV from Linac 2 to PSB and accelerated to 1.4 GeV. Continuing their journey through the PS and SPS, the protons eventually are accelerated up to 7000 GeV in the LHC and lead to interaction. Except the Linac 2, which is a linear particle accelerator, the other four accelerators are all synchrotrons.

An accelerator has four main components:
- a vacuum tube, in which the particles are accelerated and collided
- dipole magnets to keep the beam of particles in the circle orbit
- quadrupole magnets to focus the beam of particles
- electromagnetic resonators to accelerate the particles and compensate for their energy losses

The LHC itself is not a perfect circle of 26.7 km, but rather an octant. It has 8 arcs and 8 insertions. The arcs are the magnetic parts of the LHC, that bend the particle beam into a circular shape. The insertions are linear parts, in which the particles are accelerated and also used for physical experiments or maintainance of the particle beam. Between the arcs and the insertions there are the 8 transition regions.

To detect the products of a collision event detectors have to be used, the data then has to be stored and processed by computers. Given that the LHC can generate about 600 million particle collisons per second, an enormous amount of data is created, processed and stored in a very short amount of time.

A detector's purpose is to count, track and characterise all the products of a collison, so that a reconstruction of the collision is possible. The general particle detector consists of four main parts:
- Tracker: aimed to track charged particles
- Electromagnetic calorimeter: aimed to catch photons and electrons
- Hadron calorimeter: aimed to catch hadrons
- Muon spectrometer: aimed to detect muons by tracking their specific path through the spectrometer

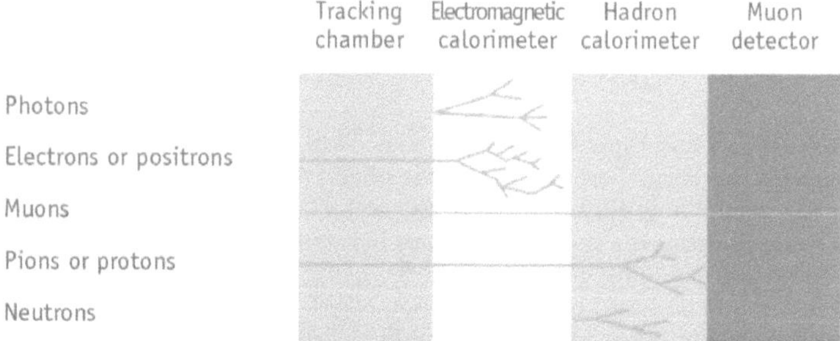

Figure 10: Scheme of a particle detector

There are 4 main and large detectors that should be mentioned briefly. Each of them occupies one of the 8 insertions available at LHC.
- ALICE: detects lead-ion collision and studies the properties of quark-gluon plasma
- ATLAS: is a general-purpose detector and covers a wide range of purposes including the search for the Higgs boson, supersymmetry and extra dimensions
- CMS: is a general-purpose detector and covers the same purposes as ATLAS, but is built differently from ATLAS and therefore provides an independent measurement from ATLAS
- LHCb: detects particles based on the bottom quark and studies the asymmetry between matter and antimatter in the universe

The next section will focus on the Experimental Data from ATLAS and CMS, that lead to the discovery of the Higgs boson. More details can be found in CERN's LHC brochure.[36]

4.3 Experimental Data from ATLAS and CMS

Results of an analysis based on collision data collected from the LHC at 7/8 TeV in 2012 suggests the discovery of a new neutral boson of mass ~125 GeV. The two independently working collaborations ATLAS[28] and CMS[29] have both found a significant excess of events of 5.9σ standard deviation at around 124-6 GeV, indicating the creation of a new particle during the collision process.

The exact result from ATLAS is 126.0 $\pm$ 0.4 (stat) $\pm$ 0.4 (sys) GeV, compared to the result 125.3 $\pm$ 0.4 (stat) $\pm$ 0.5 (sys) GeV of CMS. The results are given with a statistical error, which arrises from the analysis of the data and a systematic error, that is the uncertainty of the measurements from the equipment.

Both experiments focus on the five channels that are described in section 3.2. The general process is to measure all events of collisions, to take data of selected events and collision products. Subsequently, to estimate the background processes by computer simulations, all processes that could be like a Higgs decay, but are not. Finally, to compare the estimated background with the actual measured results. If there is an excess in events, that could suggest new particles. This excess was measured and announced in July 2012.

The 5 channels are divided in higher mass resolution channels: $H \to ZZ$, $H \to \gamma\gamma$ and lower mass resolution channels: $H \to WW$, $H \to \tau\tau$, $H \to bb$

Results of the higher mass resolution channels of CMS and ATLAS will be presented in the following.

1. $H \to ZZ \to 4\,l$

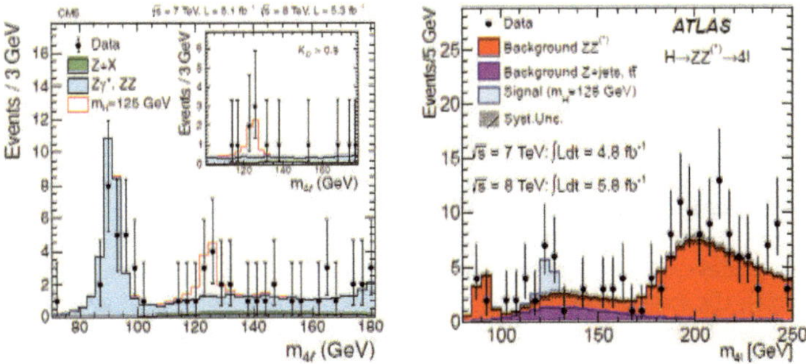

Figure 11: Distribution of the 4-lepton invariant mass, comparison of the experimental data with the estimated background and the expected contribution of a Higgs boson at 125 GeV, left: CMS, right: ATLAS

The background estimation arrises from the processes of Z+jets and $t\bar{t}$, ZZ*, $Z\gamma$ and Z+X, with the dominant process being $q\bar{q} \rightarrow ZZ^*/Z\gamma$ This process is estimated by a Monte Carlo computer simulation using the transition probabilites given by the Standard Model. The background Z+jets and Z+X are estimated based on experimental data. The process $t\bar{t} \rightarrow W^+b + W^-\bar{b}$ with the W decaying into 2 leptons is also estimated via computer simulation.

Comparing the experimental data from the collisons at LHC, it can be seen that there is an excess of events at around 125 GeV, that can be explained by the decay of a Higgs boson into ZZ and subsequently into 4 leptons.

2. $H \rightarrow \gamma\gamma$

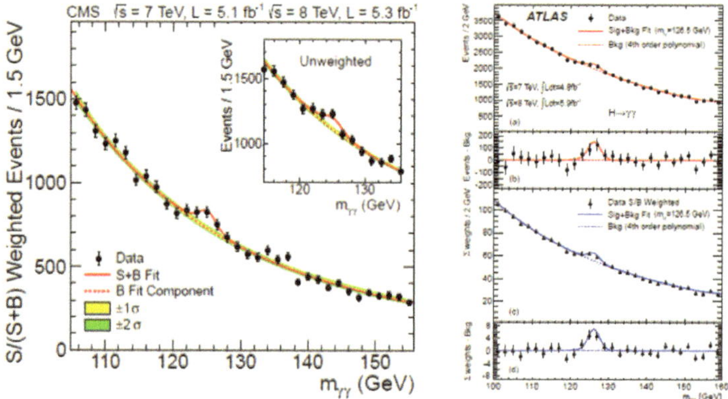

Figure 12: Distribution of the di-photon invariant mass, comparison of the experimental data with the estimated background and the additional amount of events measured at 126 GeV, left: CMS, right: ATLAS

The background is estimated by data on diphoton invariant mass spectrum fitted into a polynomial model. The order is typically chosen from 3 to 5, in order to achieve a compromise between statisical resolve and potential bias.

Comparing the experimental data with the background and signal a little excess of events can be seen at around 126 GeV. The weighted/unweighted background and signal events are fitted into a line and compared to the polynomial model. The ATLAS figure also shows the residual of the excess.

These additional events can be explained by a Higgs boson of mass 126 GeV decaying into 2 photons.

Results from ATLAS[30] and CMS[31] presented in March 2013 consolidate the assumption that the new boson is indeed a Higgs boson of mass $\sim$125.4 GeV.

5 Conclusion

The Higgs mechanism, that enables a spontaneously broken symmetry, can be summarised the last two parts of (1). Which is adding a new field, the Higgs field h, to the Lagrangian of the Standard Model. Its dynamical term $|D_\mu\Phi|^2$ interacts with fermions and bosons, thus giving them a mass, its potential term $V(\Phi)$ gives the mass term for the Higgs boson $m_H = \sqrt{\frac{\lambda}{2}}v$. Theoretical constraints on the mass of the Higgs boson can be obtained by means of the pertubation theory.

Experimental search for the Higgs boson has concentrated on the range 115-135 GeV at CERN's LHC experiments ATLAS and CMS. A boson most likely the Higgs boson of the Standard Model has been found to have a mass of around $\sim$126 GeV in July 2012. That completes the search for the Higgs boson and a formal recognition of its existence was given on the 10th of December 2013 by the Nobel committee by awarding the Nobel Prize in Physics 2013 to F. Englert and P. Higgs, who "have [...] found the key to understanding the masses of elementary particles"[35].

Since the existence of a Higgs boson has been confirmed and the Standard Model(SM) is to some extent complete, the focus goes further and beyond. Other models and extensions have been theoretically worked out and experimental search for evidence is ongoing. One popular theory is called Supersymmetry(SUSY). The most simple addition to the SM is the Minimal Supersymmetric Model (MSSM). It adds another complex douplet of a Higgs field to stabilise the mass of the Higgs bosons. That will result in 5 Higgs bosons: three neutral ones h, H, A and a couple of charged ones $H^{\pm}$. Little h can be seen as the found "SM-like" Higgs boson, capital H as a "SM-like" Higgs boson but heavier, A is a superysmmetric addition with different properties than the SM-Higgs boson and $H^{\pm}$ as a pair to conserve charge in interactions with other charged elementary particles. For a detailed review on MSSM, see references[37][38].

The search for "the Higgs boson" is still not over yet. It might lead us to new physics beyond the standard model.

6 Bibliography

References

[1] C. N. Yang, R. Mills (1954): "Conservation of Isotopic Spin and Isotopic Gauge Invariance", Physical Review 96: 191-195, Bibcode:1954PhRv...96..191Y, doi:10.1103/PhysRev.96.191

[2] S. Glashow (1961): "Partial-symmetries of weak interactions", Nuclear Physics 22: 579-588, Bibcode:1961NucPh..22..579G, doi:10.1016/0029-5582(61)90469-2

[3] F. Englert and R. Brout (1964): "Broken Symmetry and the Mass of Gauge Vector Mesons", Physical Review Letters 13: 321-323, Bibcode:1964PhRvL..13..321E, doi:10.1103/PhysRevLett.13.321

[4] P. Higgs (1964): "Broken Symmetries and the Masses of Gauge Bosons", Physical Review Letters 13: 508-509, Bibcode:1964PhRvL..13..508H, doi:10.1103/PhysRevLett.13.508

[5] S. Weinberg (1967): "A Model of Leptons", Physical Review Letters 19: 1264-1266, Bibcode:1967PhRvL..19.1264W, doi:10.1103/PhysRevLett.19.1264

[6] A. Salam, N. Svartholm, ed. (1968): "Elementary Particle Physics: Relativistic Groups and Analyticity", Eighth Nobel Symposium, Stockholm: Almquvist and Wiksell, p. 367

[7] R. Mann (2010): "An Introduction to Particle Physics and the Standard Model", CRC Press Taylor & Francis Group

[8] http://upload.wikimedia.org/wikipedia/commons/0/00/ Standard_Model_of_Elementary_Particles.svg

[9] J. Shifflett (2013): "Standard Model Lagrangian (including neutrino mass terms)", Cambridge University Press, Cambridge, 2007, http://einstein-schrodinger.com/Standard_Model.pdf

[10] W. Miller (1972): "Symmetry Groups and Their Applications", Academic Press, New York and London, Volume 50 in PURE AND APPLIED MATHEMATICS, p. 13, p. 173-175

[11] M. Tavel (1971): "Invariant Variation Problems", English translation of E. Noether (1918): "Invariante Variationsprobleme", http://arxiv.org/abs/physics/0503066

[12] K. Brading, H. Brown (2000): "Noether's Theorems and Gauge Symmetries", http://arxiv.org/abs/hep-th/0009058

[13] S. Dawson (1999): "Introduction to Electroweak Symmetry Breaking", http://arxiv.org/abs/hep-ph/9901280

[14] K. Nguyen (2009): " The Higgs mechanism", http://www.theorie.physik.uni-muenchen.de/lsfrey/teaching/archiv/sose_09/ rng/higgs_mechanism.pdf

[15] I. Aitchison, A. Hey (2004): "Gauge Theories in Particle Physics, Volume II: QCD and the Electroweak Theory", Institute of Physics Publishing, p. 211-215, p. 269-271, p. 308-388

[16] http://www-zeus.physik.uni-bonn.de/ brock/feynman/vtp_ws0506/ chapter09/higgs_pot.jpg

[17] B. de Wit (2011), Lectures Notes on Field Theory in Particle Physics 2011, chapter 11, http://www.staff.science.uu.nl/ wit00103/ftip/Ch11.pdf

[18] J. Goldstone (1961): "Field Theories with Superconductor Solutions", Nuovo Cimento 19: 154164, doi:10.1007/BF02812722

[19] L. Álvarez-Gaumé, J. Ellis (2011): "Eyes on a prize particle", Nature Physics 7, 2/3, doi:10.1038/nphys1874, http://www.nature.com/nphys/journal/v7/n1/fig_tab/nphys1874_F1.html

[20] Particle Data Group (2012): "Particle Physics Booklet", http://pdg.lbl.gov/2013/download/rpp-2012-booklet.pdf

[21] G. Bernardi, M. Carena, T. Junk (2012): "HIGGS BOSONS: THEORIES AND SEARCHES", Particle Data Group, http://pdg.lbl.gov/2013/reviews/rpp2012-rev-higgs-boson.pdf

[22] M. Spira, A. Djouadi, D. Graudenz, P.M. Zerwas (1995): "HIGGS BOSON PRODUCTION AT THE LHC", http://arxiv.org/abs/hep-ph/9504378

[23] A. Denner, S. Heinemeyer, I. Puljak, D. Rebuzzi, M. Spira (2011): "Standard Model Higgs-Boson Branching Ratios with Uncertainties", http://arxiv.org/abs/1107.5909

[24] G. Degrassi, S. Di Vita, J. Elias-Mir, J. Espinosa, G. Giudice, G. Isidori, A. Strumia (2012): "Higgs mass and vacuum stability in the Standard Model at NNLO", http://arxiv.org/abs/1205.6497

[25] A. Wingerter (2011): "Implications of the Stability and Triviality Bounds on the Standard Model with Three and Four Chiral Generations", http://arxiv.org/abs/1109.5140

[26] LEP (2011): "Search for the Standard Model Higgs Boson at LEP", http://lephiggs.web.cern.ch/LEPHIGGS/papers/LEP-SM-HIGGS-PAPER/index.html

[27] Fermilab Tevatron (2012): "Updated Combination of CDF and D0 Searches for Standard Model Higgs Boson Production with up to 10.0 fb-1 of Data", http://tevnphwg.fnal.gov/results/SM_Higgs_Summer_12/, http://arxiv.org/abs/1207.0449

[28] ATLAS (2012): "Observation of a new particle in the search for the Standard Model Higgs boson with the ATLAS detector at the LHC", http://arxiv.org/abs/1207.7214

[29] CMS (2012): "Observation of a new boson at a mass of 125 GeV with the CMS experiment at the LHC", http://arxiv.org/abs/1207.7235

[30] ATLAS (2013): "Combined measurements of the mass and signal strength of the Higgs-like boson with the ATLAS detector using up to 25 fb-1 of proton-proton collision data", http://cds.cern.ch/record/1523727

[31] CMS (2013): "Observation of a new boson with mass near 125 GeV in pp collisions at $\sqrt{s} = 7$ and 8 TeV", http://arxiv.org/abs/1303.4571

[32] ATLAS (2013): "Evidence for the spin-0 nature of the Higgs boson using ATLAS data", http://arxiv.org/abs/1307.1432

[33] ATLAS (2013): "Measurements of Higgs boson production and couplings in diboson final states with the ATLAS detector at the LHC", http://arxiv.org/abs/1307.1427

[34] ATLAS and CMS collaborations (2013): "Birth of a Higgs boson", CERN Courier, http://cerncourier.com/cws/article/cern/53086

[35] Nobel Media AB (2013): "The Nobel Prize in Physics 2013", http://www.nobelprize.org/nobel_prizes/physics/laureates/2013/, http://www.nobelprize.org/nobel_prizes/physics/laureates/2013/presentation-speech.html

[36] CERN LHC (2009): "CERN faq LHC the guide", http://cds.cern.ch/record/1165534/files/CERN-Brochure-2009-003-Eng.pdf

[37] J. Gunion, H. Haber (1984): "Higgs Boson in Supersymmetric Models", SLAC-PUB-3404, Stanford University, http://www.slac.stanford.edu/cgi-wrap/getdoc/slac-pub-3404.pdf

[38] M. Carena et al. (2013): "MSSM Higgs Boson Searches at the LHC: Benchmark Scenarios after the Discovery of a Higgs-like Particle", http://arxiv.org/abs/1302.7033